HOT SPOTS

H O T S

A M E R I C A ' S V O L C

B L A C K - A N D - W H I T E P H O T

C O L O R P H O T O G R A P H

W I T H A N I N T R

G A R R E T T

A B U L F I N C H P R E S S B O O K

B O S T O N · N E W Y O R K

P O T S

A N I C L A N D S C A P E

O G R A P H S B Y D I A N E C O O K

S B Y L E N J E N S H E L

D U C T I O N B Y

H O N G O

L I T T L E, B R O W N & C O M P A N Y

T O R O N T O • L O N D O N

TO OUR PARENTS, BILL AND TIL COOK, AND AL AND ANN JENSHEL.

 First Edition. All photographs courtesy of Yancey Richardson Gallery. Library of Congress Cataloging-in-Publication Data. Cook, Diane and Jenshel, Len. Hot spots: America's volcanic landscape/black-and-white photographs by Diane Cook; color photographs by Len Jenshel; introduction by Garrett Hongo.—1st ed. p. cm. "A Bulfinch Press book." ISBN 0-8212-2266-X 1.Volcanoes—United States—Pictorial works. 2.Geothermal resources—United States—Pictorial works. I. Cook, Diane and Jenshel, Len. II. Title. QE524.C66 1996 551.2'1'0973—DC20 96-1782. Bulfinch Press is an imprint and trademark of Little, Brown and Company (Inc.) Published simultaneously in Canada by Little, Brown & Company (Canada) Limited. Book Design: Louise Fili Ltd. Design Assistant: Tonya Hudson. PRINTED IN ITALY.

1/99

CONTENTS

THERE IS NO SCIENCE WITHOUT FANCY,
AND NO ART WITHOUT FACTS.

VLADIMIR NABOKOV

PREFACE

THE FIRST TIME WE SAW LAVA FLOWING OUT OF THE EARTH WE WERE HOOKED. Surrounding us was a stark, paradoxical landscape that spoke simultaneously of creation and destruction. It was a vision that might prompt some to flee, but we found ourselves drawn instead by its primordial power. Since that transforming moment on Kīlauea Volcano in 1991, we have photographed close to one hundred hot spots, volcanoes, and their progeny —cinder cones, lava tubes, and geysers—across America.

That moment sealed our commitment to this project, but the idea of collaborating had occurred to us years before. In the 1980s, shortly after we were married, we were both photographing in the American Southwest. Despite our different approaches to landscape photography—Diane quicker and more intuitive, Len slower and meditative—we found ourselves attracted to similar subject matter. We discovered after emerging from our darkrooms that the photographs struck a counterpoint when viewed together. Thus began the idea of pairing.

The volcanic landscape became the specific focus of our project during a trip to Yellowstone. Our initial intent was to document the rebirth of the forest after the devastating fires of 1988. But while there, we learned that geologists had determined that Yellowstone lies within the caldera of an immense volcano. A landscape altered by a windblown fire no longer seemed as compelling as one transformed by fires deep within the earth. Where else, we wondered, had volcanism wrought its effects, subtle or dramatic, on the landscape?

The following year we traveled to the most active hot spot in the United States, Kīlauea Volcano on the Big Island of Hawai'i. We hiked out across the flanks of the volcano to the spot where molten rock poured into the sea from Pu'u 'Ō'ō's lava tube. As dusk approached we witnessed the most mesmerizing fireworks display of our lives, heralding the formation of new land. At that instant we knew our challenge would be to document the age-old and continuous process of the earth creating itself.

We approach our survey of America's volcanic landscape as artists and enthusiasts, not as scientists. All of the photographs in this book are of volcanic phenomena; however, we have taken poetic license with the term "hot spots," for though these features were once hot, they are not all technically caused by hot spots.

Though we are able to trace the origins of our endeavor, we have not yet reached its end. Our passion persists when we hear that more than twelve hundred earthquakes shook Yellowstone in 1995. Will the effects of this hidden volcano turn suddenly from gentle to cataclysmic? The question reminds us that the earth's process of creation is an ongoing one, and that volcanoes will continue to affect our landscape, our lives, and our imaginations.

DIANE COOK AND LEN JENSHEL
NEW YORK CITY, MARCH 1996

EXPLOSIONS OF LAVA, HAWAI'I VOLCANOES NATIONAL PARK, HAWAI'I, 1991

INTRODUCTION

Ministry: Homage to Kīlauea

Living on the mainland and in cities these past few years, I have grown confused about the earth. It had been my thought for a long stretch of time during my middle thirties—those years I'd lived in Volcano, Hawai'i, that village in the rain forest where I was born—that one could learn to *belong* to a place, even a chosen place, in a way that would feel like home, like an inheritance of some spiritual and nonmaterial sort; that, in time, the mere apprehension of a particular spot of earth would come to sing out like a choir of its finest birds and extend you the gift of some intimate warmth, almost like recognition, like kinship itself. It was as if the natural features in a landscape were manifestations of a sublime consciousness, that mountains and canyons, that calderas and sea arches had something indwelling and harmonious with my soul's need for rest and acceptance. Not Maia's veil, but the face of God.

Lately, though, I have felt otherwise, bereft, that these beliefs are consolational fictions. That the earth, despite any of our affections for it, is itself and spiritually inviolable, austere and removed from any human homage or protestation. It is a lordly and an inhumane magnificence. It is power and preciousness. It is itself. Dread.

When this desolate feeling comes upon me, I know I have been in cities too long. I know I need to make a small journey and gaze upon the pure creation of earth. I go walking upon volcanoes.

• • •

When my two boys were babies, to help them fall asleep in the afternoons, I liked driving them out from the house we always rented in the village of Volcano, not three miles from the summit caldera of Kīlauea. We lived in Mauna Loa Estates, a pattern of subdivided lots cut out of the rain forest downslope of the volcano. Every day around three o'clock, I would drive my kids up the highway a mile or two through the park entrance, our little car plunging like a submarine past all the microclimates and botanical realities of tree ferns and runaway exotic gingers and bamboo orchids until we got to the swing in the road just before the turnout to Kīlauea observatory at Uwēkahuna Bluff. I'd pull over into a little gravel slot by the roadside and let all the air-conditioned tour buses and shining red rental cars and USGS Cherokees and geologists' Broncos swoosh by while I took a long view over the dome of Kīlauea toward the veldt-like lower slopes of Mauna Loa, my boys already asleep in the backseat. It became part of a contemplative ritual for me—the afternoon drive with my children, then the pure gazing into a spectacular landscape. What I liked was the swoop of land, the way it rolled out from under my beach-sandaled feet, and the swimming air, freighted with clouds that seemed the land's vision rising over it. I always felt that I could have been the land's own dream then, and I liked thinking of myself that way, as offspring come to pay it the tribute of my own assembling thoughts seeking the unearned coherence of apprehending beauty, little brainy cyclones that touched down in the lava channels or drained back into rivulets of wind. "Cloud and Man differ not," I'd joke to myself. "All is One under Heaven." And why not?

I thought in romantic rhythms; I thought: *What if we were to recast ourselves as descendants all gathered at the foot of our sponsoring mountains, drawn by a love like primitive magnetisms and convection currents calling all things back to their incarnate sources? What would be the point other than to step into the sulfuric cleansing of volcanic clouds? Our dithyrambs of dream-mountains not quite earth's equal but more vague than that—like clouds around Mauna Loa, drifting continents of vapor and dust riding the gyring wind-gusts over Halema'uma'u and Iki, mantlings of evanescence on the tropical shoulders of an angel? Aren't we the earth become known to itself, I thought, Homo sapiens celebrants of a sublime not completely dreadful, but companionable too, its presence like two sleeping children, innocent dragons fogging the car's rear window with a visible breath?*

• • •

A few years ago, a friend from the Mainland, a poet and literary critic, flew to visit me in my remote seclusion from the literary world. My friend is immersed in the fray of the *New York Times Book Review*, *Poetry*, and *The New Yorker*, and I value his insights and opinions. In the few days of his stay, I took him to lookouts and photo opportunities around the volcano while we schmoozed on about editors and poetry readings, about book contracts and creative writing programs, about museum shows and the NBA playoffs.

We circumnavigated the caldera counterclockwise along the road named Crater Rim Drive. I spouted on about the cloud-magnet that Mauna Loa was, how it gathered its weather every day, wearing a skirt of clouds at about the sixty-five-hundred-foot level by the time three o'clock rolled around. I stopped at various lookout points: a little roadside rise near the volcano observatory inside the national park; a pullout overlooking the hogback-shaped fumaroles of what was once a "curtain-of-fire" eruption along Kīlauea's northern caldera floor; a turnout next to Keanakakoi, a smallish pit crater where we could see the ragged earth drop six hundred feet to an almost smooth black lava floor; the viewpoint over Iki where I'd once seen schoolchildren on a field trip and heard them singing as they crossed the caldera; the still, quiet arena in the rain forest near there where the tree line grew a little higher and the tree ferns and false staghorns thickened over patchy hot spots still heating through the undergrowth.

After a couple of days of this, I realized I was turning botanical and geologic wonders into patter, spiel, and personal history, giving my friend a kind of live slideshow presentation of the volcano. As an antidote, I decided we should hike across the short, flat trail from a turnout along the Chain of Craters Road toward a dormant lava cone, centuries old, named Pu'u Huluhulu. I wanted to show him the lava trees along the way. They were the remains of a forest burned down by one of the recent flows during the seventies.

Lava trees form in the aftermath of the movement of a large flow of *pāhoehoe*, that viscous lava, through a stand of *'ōhi'a* or *hāpu'u*, the native myrtle and gigantic tree fern of Hawai'i. *Pāhoehoe* streams through the grove of trees, incinerating bark and branches, but flowing around the fatter trunks, cooling down, forming lava casts around the burning trees as the flow ebbs away.

A little path ran *makai* (seaward) of the remnant grove we walked through. Huge geoducks and horse penises of black spinnakered glass stood out from the flow. These were forbidding snags formed around tree ferns and *'ōhi'a*—spires of *pāhoehoe* left behind while the flow had driven on. Encased in lava, the trunk woods incinerated and burned, leaving centers filled with charcoal and ash, slowly hollowing out in every breath of fresh wind. When everything cooled and became solid, what was left was a Grove of Suicides—Dante's fiery souls trapped in the remains of trees that bled black rock at their joints and seams, that shattered and flaked when you touched them, that, from a distant view, made a plain of charred stubble under dark, drenching skies. I wrapped an arm around one and, though I felt life and not death there, the rock tree more a shriveled and torn umbilical than the condemned form of one of Dante's sinners, I sensed how another could imagine an ending coming to the earth and our time on it, a forest of green sacrificed to this black of a new beginning, to me a froth of redemption.

My friend snapped photos for a while, and I sensed the visit was becoming mechanical. He was at a loss out there on the black expanse of lava. It was the moment of turning as with any freighted experience or relationship. A line of activity exhausts itself and must change. The fires of that wild ecstasy for the world and poetry we had shared in our youth had banked some, and we were moving away from each other.

He gestured to a bomb of rock near me. His face had brightened and the cloud of emotions or boredom had cleared. He picked up the fractured white limb of an *ʻōhiʻa* and swung it like a baseball bat. He wanted a pitch, a game. I was perplexed, but I picked up a piece of rock and lobbed it like a softball, low and away. Ready for it, he took a swing and missed, laughing.

I thought of our friendship, how it was built on the way we each cherished the particular histories of the other's family, how both of us, during our college years and years of apprenticeship in the art, had taught ourselves ways to love the world through reading its many poets—Neruda of Chile, Li Po and Tu Fu of China, Whitman of Manhattan and Brooklyn and New Jersey, Yeats of Ireland. Yet I was moving away from books now and into something else I thought might eventually mean more—I felt I was moving toward the land. Perhaps even *my* land. It felt odd to me to be playing this boyhood game out on the plain of lava trees. But I went along, tossing the different pieces of lava I could find, trying to recapture a feeling for each other. It all felt so dissonant to me, though I tried to lighten my heart to match his frivolity, getting into the sport and mock competition of it all.

"That's a double," he'd say, shattering a slab of frothy rock and sending its largest fragment over my head and fifteen yards over the slick surface of the hardened flow.

"That's a pop-up," I'd say, or, "Strike three, side retired." White spider chrysanthemums of clouds opened against the blue azimuth of sky overhead. The concussive *whup-whup* of a helicopter went off far away.

The first time I'd visited Volcano, I was filled with the immensities of realization and resolve. It was like an eclipse—the wolf of the moon devouring the sun. Yet it could not captivate my friend, and all my blather about it seemed only to have gotten in the way. He was not as susceptible to this volcanic beauty. I knew that in his poetry he took inspiration from paintings and modern music, from life in American cities of the East and Midwest, and from poets of the European past. His concern was the desolation in human spirit after the fall of civilization, yet again, even after the brilliant optimism of our America. We were brothers escaping the old literary myths, living in passion and different histories, poets of sentiment improvising a way to speak and to act while living on the earth.

At play on that lava field, we began talking about what fuels creative acts, how we'd each found our own ways to make poems. I bent down and took a pinch of brown silica, called Pele's hair, between my fingers. There was amber and viridian and bronze all apparent in the fine fibers of rock. I tossed them away and told him that the wish to make poetry had begun as an ineffable *feeling* for me, one I didn't necessarily have words for. I tried to take it and let it lap up inside of me, a hollow chamber of being, until it filled the imagination the way lava plumes upward from its reservoir in inner earth, spouting to the surface in gouts of red display. I said I had to stay quiet, until I could see clear through a feeling that has made itself. I took inspiration from that feeling and tried to let words come from that in whatever shape they formed themselves—blossoms or fire-fountains.

"And the problem of making?" he said, excited now. "Didn't you try to shape the poem somehow? Isn't form mostly a way to build ceremony into the act of making words?"

"Form is helpful," I said, "but do we ever start with that? I always need that voice or an image that beguiles the mind, and then I wait for a *mood* to grow around that voice or that image. I try to listen to that voice, then *speak* in that voice, and I try not to direct it too much. I *listen* and I try to live in feeling."

"But how do you do that?" my friend asked. His face had wrapped itself into a thoughtful scowl. "Doesn't form and what traditions have taught us give us the tools to *stay* in the feeling, even perfect it? Can't form be a meditative tool? A way to catch the heat of a falling star?"

I saw that the far horizon had become whitened with fog behind his shoulders, that gray seas of lava spread just below the sky, that a radiance of black mounds and furled rock surrounded us.

"I try not to be too busy," I said. "Being involved with the world hurts the feeling."

Standing among the lava trees near Pu'u Huluhulu, I felt a little guilty—that I was disloyal to the land somehow. I worried that we were playing out the rules of a nineteenth-century summer game and not finding a way to respond with fitting ceremony to the stark landscape before us. My friend tossed the lava rocks toward me, and I swung the burnt limb of a tree. My ghosts and sponsoring gods were not his.

Yet I pointed to the buff-colored lava hill nearby.

"See that old cinder cone?" I said. "Doesn't it give you a feeling that's not in words?"

He turned and scanned the black sea of *pāhoehoe* around us. Lava trees stippled the plain in the distance. The sky shaped itself into an azure bowl of pure peace.

"Consciousness is the lotus dropped in a bowl," I said.

I sensed wraiths curling in the fumaroles and in the mists rising from the line of lava trees farther toward the live vent of Kīlauea. A lavish, druglike melancholy crept into my blood. Out of my body, from the gigantic lens of the sky, I could gaze down and see us standing on the little plain of black jars by the lava hill of Pu'u Huluhulu.

My friend indulged me and took in that black expanse of looking too. He was gazing off into the middle distance down the big volcanic rift in the earth. Beyond the fog line, from hill to crater and spatter cone to caldera, there was a dancer spinning out from her wrap of living rock, twirling seaward in red veils.

"Maybe ceremony could be the earth itself," he said, his face still away from me. "And words our mortal acts upon it—little shards of glory."

He turned and wheeled back to me then, suddenly, twirling his body and catching me off guard. An ink-dark piece of lava zinged past my chest. I heard it cut the air.

"Strike two!" he shouted, laughing again. My elder, gifted with a lulling sympathy and tricks of beguilement.

Before I could complain, he twirled once more and pitched me a soft, easy toss. I took a breath, then made a slow swing that connected. The splinter of lava, black as asphalt, skipped along the silver shine on the old flow's surface. We called it on that.

When we walked away, back toward the car before sunset, the black trees at our backs stood like sentinels of the afterlife, dread outfielders spaced over an earth before there was grass.

• • •

I've been looking at Kīlauea, the active volcano on the island of Hawai'i, and its various eruptive features for a few years now, and every time I do it, I really never know what it is I'll be looking at, looking for, remembering, or comparing it to. It's kind of like daydreaming, gazing at the birth-stem of all things.

This looking is a rapturous privacy and it follows whatever are the specifics of the earthly phenomenon that is before me—a fissure line of rupture in the earth fuming with sulfurous air; a glistening beach of newborn black sand; a conical driblet spire crowning a fresh flow that, out of its blowhole, spouts an incandescent emission like red sperm over the new land; solidified eddies of *pāhoehoe* swirled like fans of pandanus leaves inundating Highway 130 near Kaimū; or a frozen cascade of lava sluiced over a low dun-colored bluff that foregrounds a deep-focus panorama over the sublime shades of gray and black plain of Ka'u Desert, the mother's breast of my universe.

When I first came upon hot spots and volcanoes—whether it was the soft buff-colored lava domes of Mammoth Lakes in the eastern Sierra Nevada, the boiling earthy soups of foamy hot springs in Yellowstone Park, or the long, whale-like profile of Hawai'i's Mauna Loa—I became aware that I was before a deep mystery that shocked and yet seemed to work subtly too, driving through the green thigh of imagination, reminding me, as my thoughts formed themselves into faintly tidal rhythms of realization and befuddlement, of what was quintessential mystery—questions of poetry and creation. My mind and soul turned in a dance, squiring each other, revolving like twin stars in a galaxy of lavas flowing in a slow, mortal spiral down the bole of a frightened tree standing in its path, its crown a fan of gray coral, creating a form out of fire and extinguishment, a frothy, snakelike cone of astral matter descending the blazing trunk of an *'ōhi'a* tree burning to ash and char.

I heard a song inside of me. I had a little vision of schoolchildren crossing not a village green in London or gamboling through a countryside dotted with barns and farm ponds, but over an Alaskan plain of smokes, over a dire lake of black lavas transformed into the congenial earth they could walk upon. There simply was everything to think about, and my mind, in those moments, seemed suddenly capable of thinking them all, holding them in one breath's time, the ten thousand things of creation's entirety caught in the spirit's dance of my apprehending. A body made electric by them. It was an ocean of thoughts broader than the dark and immense plain of lavas I crossed one moonlit night driving through northern Idaho. If I know immensity, it is that piece of earth that taught it to me. *Craters of the Moon* means a dread expansiveness; means *I will be hours with one thought, one sensation;* means a reckoning of language to pure creation. All of learning, faith, and human effort can be subsumed in that momentary awareness.

Living near Kīlauea, the looking gives me details that, in the mind, pile against each other like clouds against Mauna Loa, subductions and effervescences shoring like seas against a continent. Recollecting takes meditation, another daydream, and then something like a vision comes:

If I drove up from the rain forest and through the park, plunging past all the microclimates of rain-beaded ginger lilies and the scorched forest of *'ōhi'a* and false staghorn ferns dying in the suffocating sulfuric fogs of Steaming Bluffs, crossing the sinkbowl of a tiny caldera matted with ferns and sedges, I'd get to a mound of land near Uwēkahuna where I can look out over the inflatable summit dome of Kīlauea rising over a long slot of pulverized lavas, buff and brown in sunny weather. If I stopped and pulled over, if I walked out onto the roll of the land giving way to shallow faults and gullies, an inner sea of unstable rock frequented with seismic swarms and churning with a hundred rises as if the gathering pods of khaki-backed migrating whales were spuming their way from there toward Mauna Loa, I would see a long black groove that would be Volcano Highway twisting through the gully that is the seam of earthen creation where the land, churning with perceptible movement, becomes Mauna Loa, an earthform concerto building slowly through basaltic gradients of blue and gray into the summit caldera obscured by clouds, dwarfed by the inverted bowl of the afternoon sky's pale porcelain blue. I would lean against an onrush of wind scudding over the plain of lavas, sent by the flattening heel of a cloud bank, stratocumulus, insubstantial wraith skydiving in the space between heaven and Kīlauea, between one volcano and another.

And if I reached beside me, scrubby boughs of *ohelo* bushes would be cowlicking like the thick fur between the shoulders of a wolf or a bear, ruffled alert by wind or the scent of humans walking close by. Shining berries, some a deep red, others more pale, some of them spotted with blemishes, would bounce with a sugary weight against my hand as I bent to pluck them.

The land and its atmosphere will have gathered themselves into whatever act the clouds and I might bring about: a tribute of afternoon rain, garlands of purple trailing like a fringed skirt under the moving clouds, a handful of tiny fruit, unstrung, juicy pearls tossed from my hand into the scuffed canyonlands of eroded lavas reaching out to me from a sullen sleep.

I will have roused this rock into the space of my own living. It is music. It is the sweet scent of rain spuming puffed quarter notes of dust on the land.

And if I turned to leave, back toward my car parked by the side of the road, I would be through with looking, my body shuffled against the wind like a tree, red-brilliant, full of passionate blossoms too heavy for its boughs, the long mountain at my back, pure upwelling of Kīlauea in my soul, the dancer and the dance, kin to earth again.

GARRETT HONGO
EUGENE, OREGON, 1995
FOR E. H.

THE HAWAIIAN ISLANDS

PUʻU ʻŌʻŌ CONE, HAWAIʻI VOLCANOES NATIONAL PARK, HAWAIʻI, 1993

FORMER CHAIN OF CRATERS ROAD, HAWAIʻI VOLCANOES NATIONAL PARK, HAWAIʻI, 1991

PĀHOEHOE LAVA, HAWAIʻI VOLCANOES NATIONAL PARK, HAWAIʻI, 1991

SULPHUR BANKS, HAWAIʻI VOLCANOES NATIONAL PARK, HAWAIʻI, 1991

OFFERINGS, HALEMAʻUMAʻU CRATER, HAWAIʻI VOLCANOES NATIONAL PARK, HAWAIʻI, 1993

HALEAKALĀ NATIONAL PARK, MAUI, HAWAIʻI, 1991

STEAM PLUME, HAWAI'I VOLCANOES NATIONAL PARK, HAWAI'I, 1991

THURSTON LAVA TUBE, HAWAI'I VOLCANOES NATIONAL PARK, HAWAI'I, 1993

SKYLIGHT, HAWAI'I VOLCANOES NATIONAL PARK, HAWAI'I, 1993

EXPLOSIONS OF LAVA, HAWAIʻI VOLCANOES
NATIONAL PARK, HAWAIʻI, 1991

SILVERSWORD, HALEAKALĀ NATIONAL PARK, MAUI, HAWAIʻI, 1991

PĀHOEHOE LAVA, HAWAIʻI VOLCANOES NATIONAL PARK, HAWAIʻI, 1993

KALALAU VALLEY OVERLOOK, KŌKEʻE STATE PARK, KAUAʻI, HAWAIʻI, 1991

"GLORY," HALEAKALĀ NATIONAL PARK, MAUI, HAWAIʻI, 1991

CRATER RIM ROAD, HAWAIʻI VOLCANOES NATIONAL PARK, HAWAIʻI, 1993

DEVASTATION TRAIL, HAWAIʻI VOLCANOES NATIONAL PARK, HAWAIʻI, 1993

SURVIVING HOUSE, KALAPANA, HAWAIʻI, 1993

EXPLOSIONS OF LAVA, HAWAIʻI VOLCANOES NATIONAL PARK, HAWAIʻI, 1991

EXPLOSIONS OF LAVA, HAWAIʻI VOLCANOES NATIONAL PARK, HAWAIʻI, 1993

THE PACIFIC NORTHWEST AND ALASKA

MOONRISE, MOUNT SAINT HELENS NATIONAL VOLCANIC MONUMENT, WASHINGTON, 1992

DEVASTATED ZONE, MOUNT SAINT HELENS NATIONAL VOLCANIC MONUMENT, WASHINGTON, 1992

DOWNED TREES FLOATING ON SPIRIT LAKE, MOUNT SAINT HELENS NATIONAL VOLCANIC MONUMENT, WASHINGTON, 1992

MOUNT SAINT HELENS NATIONAL VOLCANIC MONUMENT, WASHINGTON, 1992

MOUNT ADAMS FROM WINDY RIDGE, MOUNT SAINT HELENS NATIONAL VOLCANIC MONUMENT, WASHINGTON, 1992

SUMMIT OF WIZARD ISLAND, CRATER LAKE NATIONAL PARK, OREGON, 1992

WIZARD ISLAND OVERLOOK, CRATER LAKE NATIONAL PARK, OREGON, 1980

"SHE WHO WATCHES," HORSETHIEF LAKE STATE PARK, WASHINGTON, 1994

COLUMNAR BASALT, LATOURELL FALLS, COLUMBIA RIVER GORGE, OREGON, 1993

LATOURELL FALLS, COLUMBIA RIVER GORGE, OREGON, 1993

TRIPLE FALLS,
COLUMBIA
RIVER GORGE,
OREGON, 1994

SOUTH FALLS,
SILVER FALLS
STATE PARK,
OREGON, 1995

MALHEUR MAAR, DIAMOND CRATERS, OREGON, 1994

JOHN DAY FOSSIL BEDS NATIONAL MONUMENT, OREGON, 1993

MOUNT WRANGELL, WRANGELL-SAINT ELIAS NATIONAL PARK, ALASKA, 1995

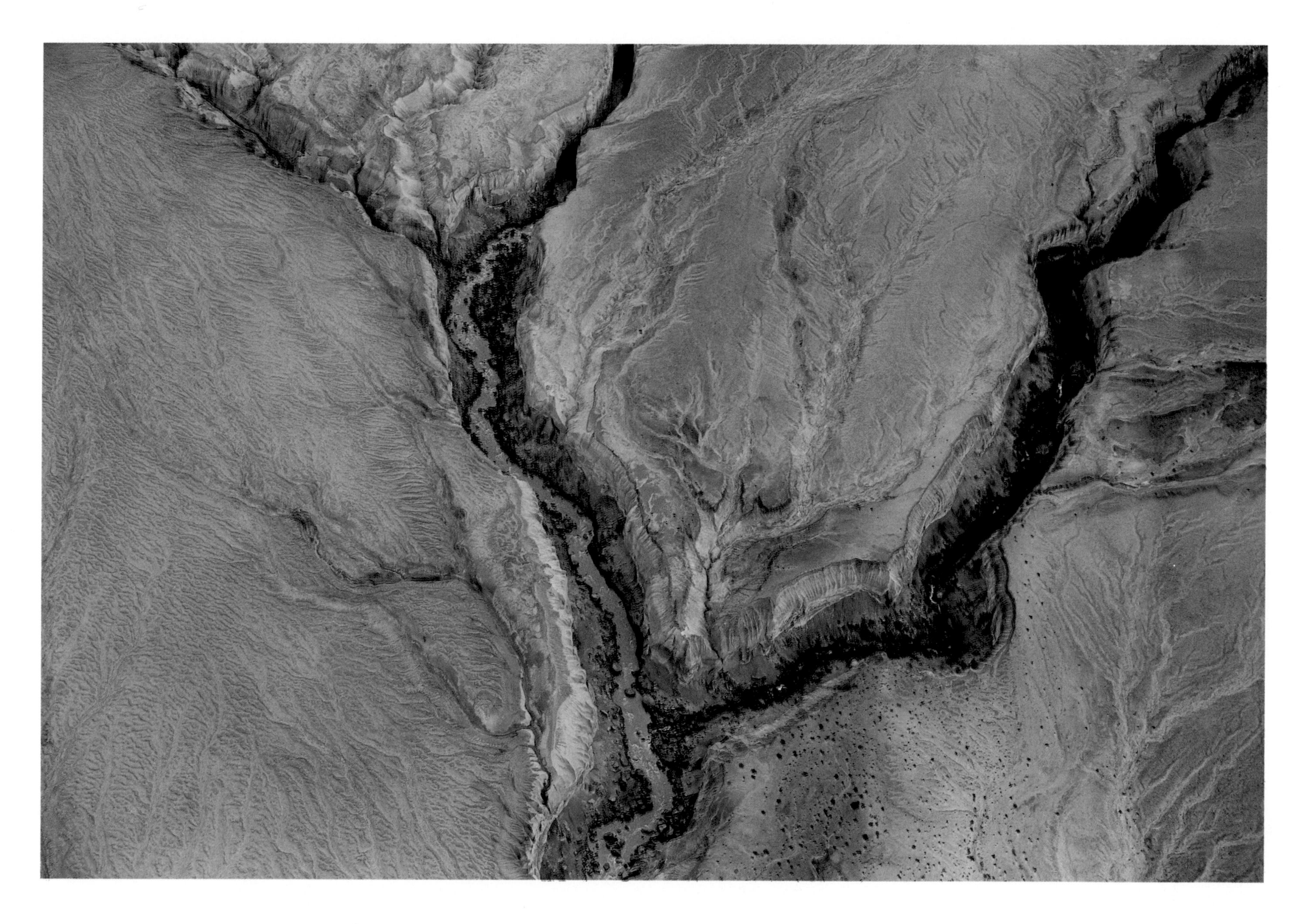

VALLEY OF TEN THOUSAND SMOKES, KATMAI NATIONAL PARK, ALASKA, 1995

MOUNT BLACKBURN, WRANGELL-SAINT ELIAS NATIONAL PARK, ALASKA, 1995

ROOT GLACIER,
WRANGELL-
SAINT ELIAS
NATIONAL
PARK, ALASKA,
1995

MOUNT BLACKBURN AND KENNICOTT GLACIER,
WRANGELL-SAINT ELIAS NATIONAL PARK, ALASKA, 1995

AUGUSTINE VOLCANO, COOK INLET, ALASKA, 1995

AUGUSTINE VOLCANO, COOK INLET, ALASKA, 1995

AUGUSTINE VOLCANO, COOK INLET, ALASKA, 1995

CALIFORNIA AND THE SOUTHWEST

SP CRATER, SAN FRANCISCO VOLCANIC FIELD, ARIZONA, 1995

GRIMES POINT ARCHAEOLOGICAL AREA, NEVADA, 1993

SHIP ROCK, NEW MEXICO, 1995

SHIP ROCK, NEW MEXICO, 1988

FERN CAVE, LAVA BEDS NATIONAL MONUMENT,
CALIFORNIA, 1994

DEVILS POSTPILE NATIONAL MONUMENT, CALIFORNIA, 1994

BLACK POINT, MONO LAKE TUFA STATE RESERVE, CALIFORNIA, 1992

APPROACHING STORM AT DAWN,
MONO LAKE TUFA STATE RESERVE, CALIFORNIA, 1991

LITTLE UBEHEBE CRATER, DEATH VALLEY NATIONAL PARK, CALIFORNIA, 1991

SNOW CANYON STATE PARK, UTAH, 1993

CITADEL RUIN, WUPATKI NATIONAL MONUMENT, ARIZONA, 1992

BENTONITE HILLS, CAPITOL REEF NATIONAL PARK, UTAH, 1988

DONEY MOUNTAIN, WUPATKI NATIONAL MONUMENT, ARIZONA, 1992

TUFF CANYON OVERLOOK, BIG BEND NATIONAL PARK, TEXAS, 1990

SUNSET CRATER NATIONAL MONUMENT, ARIZONA, 1994

BLACK BUTTE,
SHASTA-TRINITY
NATIONAL
FOREST,
CALIFORNIA,
1993

GRAND FALLS OF THE LITTLE COLORADO RIVER, ARIZONA, 1994

SUPERSTITION MOUNTAINS, LOST DUTCHMAN STATE PARK, ARIZONA, 1995

MAMMATUS CLOUDS, CINDER CONE, LASSEN VOLCANIC NATIONAL PARK, CALIFORNIA, 1992

CINDER CONE, LASSEN VOLCANIC NATIONAL PARK, CALIFORNIA, 1992

SUMMIT OF CINDER CONE, LASSEN VOLCANIC NATIONAL PARK, CALIFORNIA, 1992

YELLOWSTONE AND THE NORTHERN MOUNTAIN STATES

DEVILS TOWER NATIONAL MONUMENT, WYOMING, 1994

SPATTER CONE, CRATERS OF THE MOON NATIONAL MONUMENT, IDAHO, 1991

HOT SPRINGS STATE PARK, THERMOPOLIS, WYOMING, 1994

MINERVA SPRING, MAMMOTH HOT SPRINGS, YELLOWSTONE NATIONAL PARK, WYOMING, 1990

HOT SPRINGS STATE PARK, THERMOPOLIS, WYOMING, 1994

WHEELER GEOLOGIC AREA,
SAN JUAN MOUNTAINS, COLORADO, 1995

OPALESCENT POOL, YELLOWSTONE NATIONAL PARK, WYOMING, 1991

PUNCHBOWL SPRING RUNOFF, YELLOWSTONE NATIONAL PARK, WYOMING, 1991

PORCELAIN BASIN, YELLOWSTONE NATIONAL PARK, WYOMING, 1992

EXCELSIOR SPRING RUNOFF, YELLOWSTONE NATIONAL PARK, WYOMING, 1990

BLACK SAND POOL, YELLOWSTONE NATIONAL PARK, WYOMING, 1991

LION GROUP GEYSER ERUPTING, YELLOWSTONE NATIONAL PARK, WYOMING, 1990

PATH THROUGH PORCELAIN BASIN, YELLOWSTONE NATIONAL PARK, WYOMING, 1992

PORCELAIN BASIN, YELLOWSTONE NATIONAL PARK,
WYOMING, 1992

BACK BASIN, YELLOWSTONE NATIONAL PARK, WYOMING, 1992

ECHINUS GEYSER ERUPTING, YELLOWSTONE NATIONAL PARK, WYOMING, 1990

ECHINUS GEYSER ERUPTING, YELLOWSTONE NATIONAL PARK, WYOMING, 1990

FISHING CONE, YELLOWSTONE NATIONAL PARK, WYOMING, 1990

SELECTED VOLCANIC SITES

▲Volcanic Areas *Appears in Book

Alaska

1 Kiska Volcano
2 Kanaga Volcano
3 Seguam Volcano
4 Mount Cleveland
5 Okmok Volcano
6 Akutan Peak
7 Shishaldin Volcano
8 Pavlof Volcano
9 Mount Veniaminof
10 Aniakchak National Monument
11 Katmai National Park *
12 Augustine Volcano *
13 Lake Clark National Park
14 Mount Spurr
15 Wrangell-Saint Elias National Park *
16 Mount Edgecumbe Volcanic Field

Arizona

1 San Francisco Volcanic Field *
2 Wupatki National Monument *
3 Sunset Crater National Monument *
4 Grand Falls of the Little Colorado River *
5 Petrified Forest National Park
6 Superstition Mountains *
7 Chiricahua National Monument

California

1 Lava Beds National Monument *
2 Black Butte *
3 Mount Shasta
4 Medicine Lake Volcano
5 Lassen Volcanic National Park *
6 Clear Lake Volcanic Field
7 Mono Lake Tufa State Reserve *
8 Devils Postpile National Monument *
9 Long Valley Caldera
10 Death Valley National Park *
11 Coso Volcanic Field
12 Salton Buttes

Colorado

1 Black Canyon of the Gunnison National Monument
2 Wheeler Geologic Area *

Hawai'i

1 Na Pali Coast State Park *
2 Kōke'e State Park *
3 Diamond Head State Monument
4 Koko Crater
5 Molokini Island
6 Haleakalā National Park *
7 Kohala Volcano

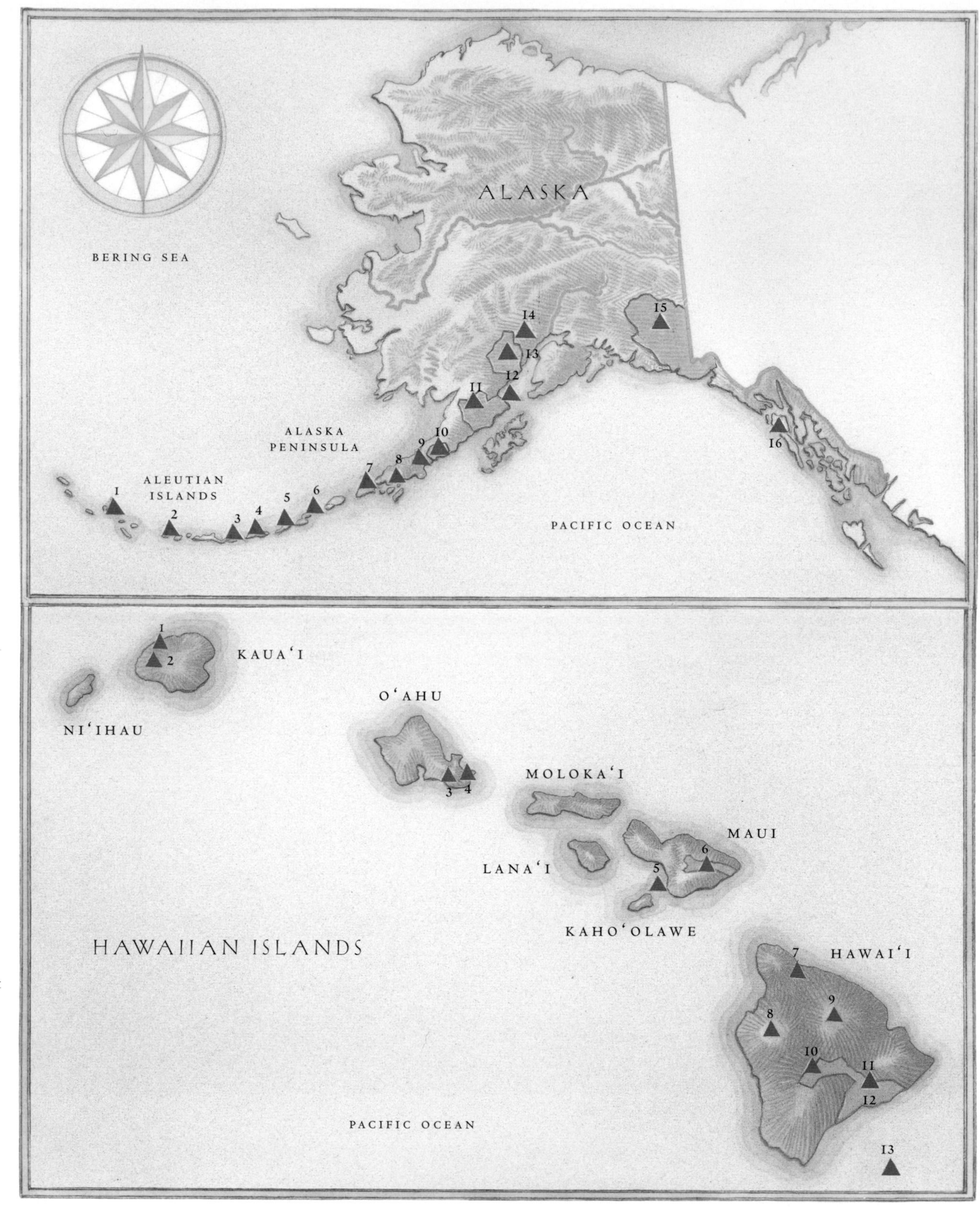

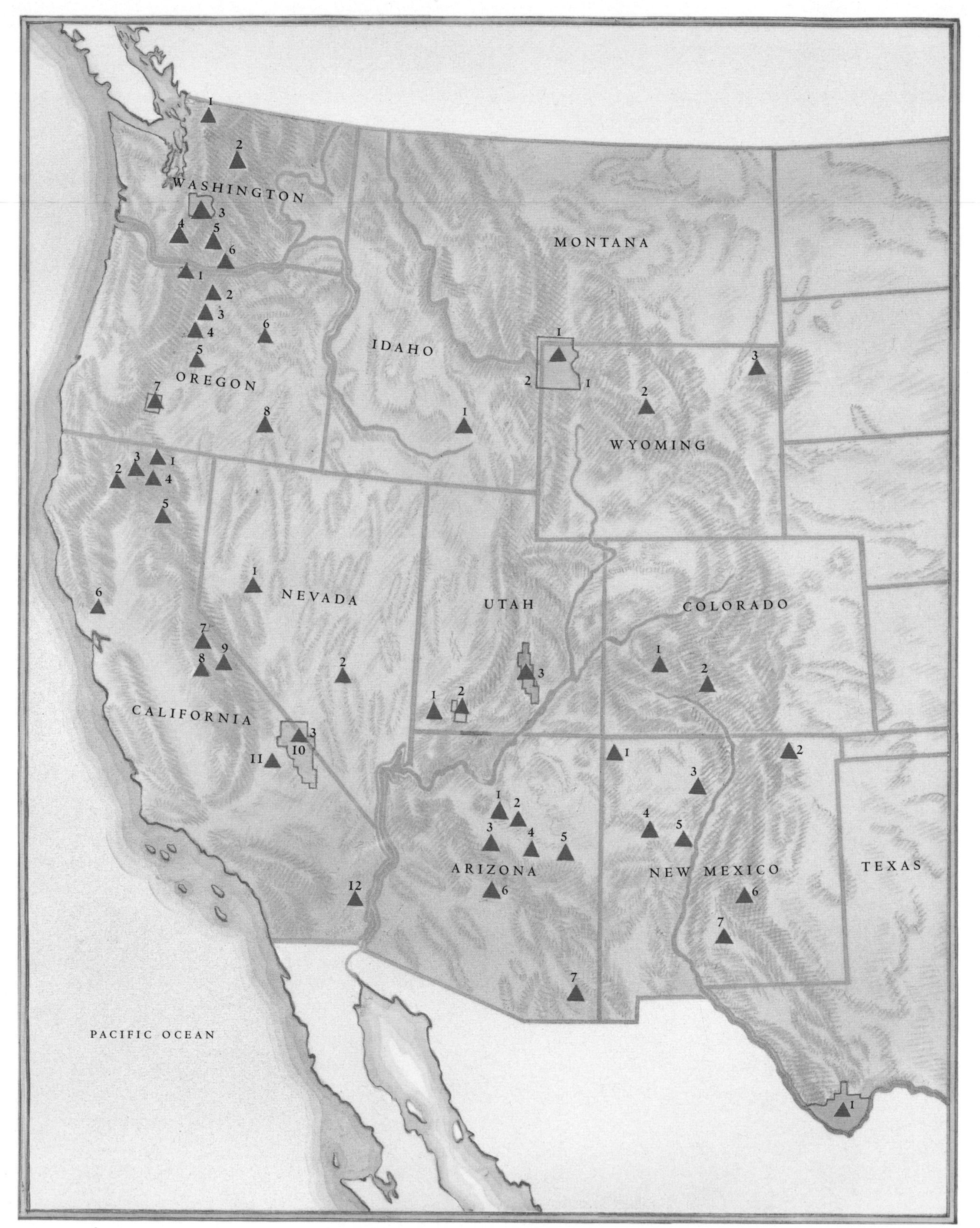

8 Hualalai Volcano
9 Mauna Kea Volcano
10 Mauna Loa Volcano
11 Kīlauea Volcano *
12 Hawai'i Volcanoes National Park *
13 Lō'ihi Seamount

Idaho

1 Craters of the Moon National Monument *
2 Yellowstone National Park *

Montana

1 Yellowstone National Park *

Nevada

1 Grimes Point Archaeological Area *
2 Lunar Crater
3 Death Valley National Park *

New Mexico

1 Ship Rock *
2 Capulin Mountain National Monument
3 Jemez Volcanic Field
4 El Malpais National Monument
5 Petroglyph National Monument
6 Valley of Fire State Park
7 Three Rivers Petroglyphs National Historic Site

Oregon

1 Columbia River Gorge *
2 Mount Hood
3 Mount Jefferson
4 Three Sisters
5 Newberry Crater National Monument
6 John Day Fossil Beds National Monument *
7 Crater Lake National Park *
8 Diamond Craters *

Texas

1 Big Bend National Park *

Utah

1 Snow Canyon State Park *
2 Zion National Park
3 Capitol Reef National Park *

Washington

1 Mount Baker
2 Glacier Peak
3 Mount Rainier National Park
4 Mount Saint Helens National Volcanic Monument *
5 Mount Adams *
6 Horsethief Lake State Park *

Wyoming

1 Yellowstone National Park *
2 Hot Springs State Park, Thermopolis *
3 Devils Tower National Monument *

The Hawaiian Islands

The Hawaiian Islands are the summits of immense volcanoes that have risen above the surface of the Pacific Ocean. They are located just about dead center in the Ring of Fire, a great circle of volcanoes stretching along the Pacific Rim to eastern Asia, all related to the shifting continental plates.

There are three types of volcanoes that relate to plate tectonics. Rift volcanoes occur when plates spread apart or separate, as in the Mid-Atlantic Ridge. Another type is caused by subduction, where one plate slides under, or subducts, another. The third kind of volcano results from a plate sliding across a stationary plume of magma, and is referred to as a hot spot volcano.

A hot spot is a persistent source of heat deep in the earth's mantle where molten rock continually rises through fractures and cracks in the crust. The Pacific plate moves at a rate of approximately four inches per year over the hot spot that has formed all the Hawaiian Islands. Each island, in a conveyor-belt fashion, is first formed and then subsequently cut off from the magma supply below, becoming dormant and finally extinct. Farthest to the southeast of the Hawaiian Islands is the newest volcanic island, known as Lō'ihi, now forming thousands of feet below sea level and expected to surface in about ten thousand years.

At present, the Big Island of Hawai'i is the eruptive attraction, with two active shield volcanoes—Mauna Loa and Kīlauea. These produce eruptions of a type known as Hawaiian eruptions—usually gentle and slow-moving lava flows. Mauna Loa's summit is 13,696 feet above sea level, not the tallest peak on earth, but when measured from the ocean floor, it is over 31,000 feet tall, making it the largest mountain in the world—and still growing! Kīlauea, perhaps the most studied volcano in the world, has been intermittently erupting since 1823. In 1983, a new vent opened and has been continuously erupting to date.

• • •

PAGE 15: PU'U 'Ō'Ō CONE,
HAWAI'I VOLCANOES NATIONAL PARK, HAWAI'I, 1993

Pu'u 'Ō'ō is a vent of the Kīlauea Volcano and, as of this writing, one of the most active volcanoes in the world. It began erupting in 1983 and continues to this day—it is the largest rift-zone eruption in recorded history. Lava mostly flows underground through a series of lava tubes (see page 22), ending seven miles away with steam and explosions as hot lava hits the ocean (see pages 21, 24–25, 34, and 35). The colored disk surrounding the shadow of the helicopter is an atmospheric phenomenon called a glory (see page 29).

PAGE 16: FORMER CHAIN OF CRATERS ROAD,
HAWAI'I VOLCANOES NATIONAL PARK, HAWAI'I, 1991

While driving down the Chain of Craters Road from the Kīlauea Volcano to the ocean, we noticed a sliver of old road partially buried by numerous lava flows. A National Park Service sign reading *alani kahiko,* which they translated as "closed road," confirmed the obvious. Between 1969 and 1974, Mauna Ulu's eruptions covered twelve miles of this section of the road.

• • •

PAGE 17: PĀHOEHOE LAVA,
HAWAI'I VOLCANOES NATIONAL PARK, HAWAI'I, 1991

This photograph was made at a pullout on the Chain of Craters Road, where a lava flow once cascaded over the cliff. Garrett Hongo has described watching an active lava flow:

> *This was called* pāhoehoe, *lava of the relatively fluid kind that congeals with a smooth and billowy surface. It cooled from its outside inward, at first forming a thin, sheetlike layer of crust on the large, flat flows. . . . When it mounded, it made a wonderland of silvers and blacks, a topiary of fantasy shapes covering the land and everything on it. I saw* pāhoehoe *that looked like the thick, fattened bodies of walruses and sea lions.* —Garrett Hongo, *Volcano*

• • •

PAGE 18: SULPHUR BANKS,
HAWAI'I VOLCANOES NATIONAL PARK, HAWAI'I, 1991

As we approached the Sulphur Banks, our noses told us that we were getting close to these fumaroles. Rainwater that seeps into cracks is heated underground and rises as steam through the vents in the Kīlauea Volcano—minerals like sulphur come along for the ride. Diane joked that her black-and-white film, still in the camera, was toning naturally.

• • •

PAGE 19: OFFERINGS, HALEMA'UMA'U CRATER,
HAWAI'I VOLCANOES NATIONAL PARK, HAWAI'I, 1993

There are various legends that describe Pele, the volcano goddess, and her travels from Polynesia to her current home in Hawai'i. In the migration legend, Pele first visits a chief on the island of Ni'ihau, where she is royally entertained. Then she happens upon a hula festival on the island of Kaua'i, where she falls in love with a handsome young chief. Determined to make a home for her new husband, she travels southeast from island to island digging pits in which she can receive her lover. It is finally on the island of Hawai'i that she successfully digs a pit without striking water, the element that extinguishes her fiery nature.

Pele still commands fear and respect. She is ill-tempered, mischievous, and subject to murderous rages. She stamps her feet to create earthquakes, summons forth lava from the bowels of the earth, and hurls lava bombs at her enemies. Scattered throughout Hawai'i Volcanoes National Park, especially along the rim of Halema'uma'u Crater, one finds offerings to the goddess of volcanoes.

• • •

PAGE 20: HALEAKALĀ NATIONAL PARK, MAUI, HAWAI'I, 1991

Haleakalā is one of the world's largest dormant volcanoes, rising over ten thousand feet above sea level. The erosional crater, obscured here by clouds, is three thousand feet deep and two and a half miles wide by seven and a half miles long. The twenty-one-square-mile crater is so large that it contains numerous climatic zones ranging from desert to rain forest. The persistent fog on the Halemau'u trail supports a thriving fern community. The new fronds of the *'ama'u* or sadleria fern turn a bright red, which protects them from the sun's harsh ultraviolet rays.

• • •

PAGE 21: STEAM PLUME, HAWAI'I VOLCANOES NATIONAL PARK, HAWAI'I, 1991

This steam plume is formed when Kīlauea's two-thousand-degree lava hits the ocean, and although it is spectacular to view, contact with it should be avoided, for it contains sulfuric acid, hydrochloric acid, and tiny shards of volcanic glass.

• • •

PAGE 22: THURSTON LAVA TUBE, HAWAI'I VOLCANOES NATIONAL PARK, HAWAI'I, 1993

A lava tube forms when a crust hardens over the surface of a *pāhoehoe* lava flow. As the outer crust cools and thickens, the core of the lava continues flowing downhill. When the fluid lava drains away, a tunnel or cave remains.

The 450-foot Thurston Lava Tube is one of the most popular tourist attractions in Hawai'i Volcanoes National Park. On any given day, buses crowd the parking lot, yet, at the extremities of the day, it can be one of the quietest and most spiritual of sanctuaries—an ironic peace evolved from the earth's violence.

• • •

PAGE 23: SKYLIGHT, HAWAI'I VOLCANOES NATIONAL PARK, HAWAI'I, 1993

It is not uncommon in the beginning stages of a Hawaiian volcanic eruption for lava fountains to shoot hundreds of feet into the air. After this episode, lava, with the help of gravity, seeks an underground "plumbing system," flowing through a series of lava tubes. When the ceiling of a tube collapses, a "skylight" allows a rare peek into a river of flowing molten rock.

One of the problems of mostly using a wide-angle lens is that you are sometimes a little too close to your subject for comfort. As we hovered over this skylight, about twenty-five feet from helicopter to opening, we were hit by an intense blast of heat from two-thousand-degree-Fahrenheit lava. But even more remarkable was the roar of the flowing lava—louder from twenty-five feet away than the rotor of the chopper a few feet above our heads.

PAGES 24–25: EXPLOSIONS OF LAVA, HAWAI'I VOLCANOES NATIONAL PARK, HAWAI'I, 1991

An hour before sunset, we arrived at the coastal site near Kamoamoa where lava ends its seven-mile journey from its source at Kūpaianaha vent and meets the ocean. During the daylight hours, the steam plume created by the molten rock hitting the ocean obscured most of the explosions of rock fragments and flowing lava. But as the sun began to set, the pyroclastic displays became more visible and mesmerizing.

The "moth to the flame" phenomenon never seemed so true as we continued getting closer to the source. You know you have gotten too close when the bottoms of your hiking boots start to melt. As John McPhee wrote in *The Control of Nature,* "Once you get red rock fever, you are never the same again."

• • •

PAGE 26: SILVERSWORD, HALEAKALĀ NATIONAL PARK, MAUI, HAWAI'I, 1991

The silversword, called *'āhinahina,* or "gray-gray," by Hawaiians, is a unique and endangered plant. It only exists in the crater of Haleakalā, with a few surviving traces on the Big Island of Hawai'i. A member of the sunflower family, the silversword has silver gray swordlike leaves that radiate from the center. Sometime between five and twenty years into the silversword's life, a center stalk shoots up as high as nine feet in just a few months' time and produces a brilliant burst of flowers. Each plant blooms just once in its life, and then dies.

• • •

PAGE 27: PĀHOEHOE LAVA, HAWAI'I VOLCANOES NATIONAL PARK, HAWAI'I, 1993

There are two different types of lava flows—*'a'ā* and *pāhoehoe.* These Hawaiian words are universally accepted as geologic terms for these flows. Although they differ in appearance, by chemical composition they are exactly the same—they differ only in the temperature of the lava and the amount of gas in the flow when the lava cools. *'A'ā* is a rough, jagged lava. *Pāhoehoe* is smooth and billowy, often with a ropy or wrinkled appearance, which occurs when a "skin" forms on the surface and is dragged along by the fluid lava underneath.

Hawaiian legend describes Pele, the volcano goddess, appearing as either a beautiful woman or an old hag. One might see her just before an eruption, in the lava fountains, or in the *pāhoehoe* flows.

• • •

PAGE 28: KALALAU VALLEY OVERLOOK, KŌKE'E STATE PARK, KAUA'I, HAWAI'I, 1991

Geologists believe that the volcanic hot spot under Kaua'i was sealed by the weight of the island itself. Mount Wai'ale'ale, the actual source of the volcano that built Kaua'i, now has the dubious distinction of being the wettest spot on earth, with an average annual rainfall of 480 inches. The Na Pali Coast, pictured here, is made up of four-thousand-foot cliffs that were carved by six million years of wind and water erosion. The lush terrain now makes it hard to imagine that this island was born of fire.

PAGE 29: "GLORY," HALEAKALĀ
NATIONAL PARK, MAUI, HAWAI'I, 1991

In Hawaiian lore, there are many interesting legends involving the mischievous demigod Maui. One, about Haleakalā, or "House of the Sun," depicts a lazy sun that so wanted to get back to bed that each day it raced across the sky in about three to four hours. Maui devised a plan to lengthen the day so that his mother would have more time to dry her tapa cloth. The demigod noticed where the first rays of the sun rose over Haleakalā Crater, so he hid in a cave at the summit and lassoed each ray as it came over the crater the next morning. With sixteen different snares trapping his rays, the sun was in no position to bargain, and so met with Maui's demand to travel more slowly across the sky.

A glory results from the fleeting combination of vapor and sun, when sunlight is diffracted toward the viewer by water droplets in a cloud. The observer's shadow is always encircled by the rainbow's rings.

• • •

PAGE 30: CRATER RIM ROAD,
HAWAI'I VOLCANOES NATIONAL PARK, HAWAI'I, 1993

Through the southwest rift zone of Kīlauea runs an earthquake fault that is Hawai'i's answer to the San Andreas Fault in California. This is one of two major fault scarps on the Big Island of Hawai'i. Rift zones are areas of weakness along a volcano that give rise to numerous flank eruptions.

In the approximately one square mile visible in this picture, one can see at least three different lava flows from the 1921, 1971, and 1974 eruptions.

• • •

PAGE 31: DEVASTATION TRAIL,
HAWAI'I VOLCANOES NATIONAL PARK, HAWAI'I, 1993

This area was created—or destroyed, depending on how you view it—by the 1959 eruption of Kīlauea Iki, or "Little Kīlauea." Lava fountains shot nineteen hundred feet into the air, showering the area with pumice, ash, and spatter. It left behind skeletons of the once vibrant *'ōhi'a* forest, a surreal tribute to the effects of volcanism.

• • •

PAGES 32–33: SURVIVING HOUSE, KALAPANA, HAWAI'I, 1993

While flying in a helicopter over a former housing development, we spotted one lone house spared the total destruction fated to the others. Fascinated by this extraordinary sight, we asked the pilot if he could fly closer. He declined because the people who lived there had been complaining about all the aircraft noise from people like us, wanting to investigate just a little bit closer. We could not believe that people still lived there, but he assured us it was true, and that they hike in a few times a week over three miles of lava with their water, propane, and food.

• • •

PAGE 34: EXPLOSIONS OF LAVA,
HAWAI'I VOLCANOES NATIONAL PARK, HAWAI'I, 1991

See note for pages 24–25.

PAGE 35: EXPLOSIONS OF LAVA,
HAWAI'I VOLCANOES NATIONAL PARK, HAWAI'I, 1993

Our second trip to the Big Island was in January of 1993, a year and a half after our first visit. We could not wait to photograph some of the things we had seen on our prior trip—like the beautiful black-sand beach at Kamoamoa; the Waha'ula Heiau, "Temple of the Red Mouth"; and the eruption site near Kupapa'u Point. In just twenty months since the first trip, the black-sand beach was gone, but the Heiau was miraculously spared. Lava still flowed underground, emptying into the ocean, but now in a different place and from a different source. Change hung unmistakably in the air—we cannot think of any other place where this phenomenon is so pronounced.

When we show this photograph, which was taken at night (by contrast, those on pages 24–25 and 34 were photographed at dusk), we are often questioned about the color of the sky—could it really have been that red? Both our memories and the film tell us that it was so. In this five-minute exposure at night, the steam plume, blurred here by movement during the long exposure, cumulatively reflects the red of the lava. In the foreground of this photograph, the world's newest black-sand beach is being formed. As of this writing, it, too, may no longer exist.

The Pacific Northwest and Alaska

Most of the Cascade Range and the Alaskan volcanoes are formed by subduction, as the Pacific Ocean plate slowly sinks beneath the western edge of North America. The plate melts when it reaches a depth of about fifty miles into the mantle, and the melted crust rises through cracks and faults in the earth's crust as magma. Generally, initial eruptions are violent and release large volumes of gas and debris. Over one hundred volcanoes and volcanic fields stretch along 1,550 miles of the Aleutian Arc in Alaska. Except for Indonesia—also part of the Ring of Fire—Alaska has more volcanic eruptions per year than any other place on earth.

• • •

PAGE 37: MOONRISE, MOUNT SAINT HELENS NATIONAL
VOLCANIC MONUMENT, WASHINGTON, 1992

On May 18, 1980, after 123 years of dormancy, Mount Saint Helens blew four hundred million tons of earth into the air with a force five hundred times stronger than that of the bomb that fell on Hiroshima.

Volcanologists have categorized three different areas to describe the effects of the blast and pyroclastic flow of the Mount Saint Helens eruption: (1) the "direct blast zone," or "devastated zone"—an area about eight miles in radius in which virtually everything was obliterated; (2) the "tree-down zone"—an area extending nineteen miles from the volcano's center in which everything was flattened; (3) the "seared zone," or "transition zone"—where trees like the ones pictured here remained standing but were scorched by the hot gases of the blast.

PAGE 38: DEVASTATED ZONE, MOUNT SAINT HELENS NATIONAL VOLCANIC MONUMENT, WASHINGTON, 1992

Geologists computed that the speed of the pyroclastic flow from the May 18 eruption of Mount Saint Helens reached 670 miles per hour, just short of the speed of sound. No scientist to our knowledge has ventured an estimate of the temperature in the direct blast zone; however, readings of 572 degrees Fahrenheit were recorded fifteen miles away from the center of the blast. Everything in that section was reduced to ash and pumice. Why this one tree stump was not obliterated like everything else in the area remains a mystery.

• • •

PAGE 39: DOWNED TREES FLOATING ON SPIRIT LAKE, MOUNT SAINT HELENS NATIONAL VOLCANIC MONUMENT, WASHINGTON, 1992

Mount Saint Helens, prior to the 1980 blast, was reputed to be the most perfectly shaped cone in the volcanic Cascade Range, often referred to as "the Fujiyama of America." The eruption blew out a crater 2,100 feet deep, dropping the summit's elevation from 9,677 feet to 8,364 feet. The avalanche of debris that tumbled into Spirit Lake raised the bottom by 295 feet and the water level by about 200 feet.

• • •

PAGES 40–41: MOUNT SAINT HELENS NATIONAL VOLCANIC MONUMENT, WASHINGTON, 1992

One sees clearly from the pattern of toppled trees in the tree-down zone the force and direction of the blast. These are the skeletal remains of Douglas fir, hemlock, and spruce that once stood over one hundred fifty feet high. Four and one half billion board feet of these giants were blown down in an area encompassing over two hundred square miles.

Standing at this site for the first time, we were simultaneously stunned, awed, and definitely humbled. Seeing these downed trees resembling toothpicks, we were most thankful that we were not standing on this spot at 8:32 A.M. on May 18, 1980.

• • •

PAGE 43: MOUNT ADAMS FROM WINDY RIDGE, MOUNT SAINT HELENS NATIONAL VOLCANIC MONUMENT, WASHINGTON, 1992

Geologists call Mount Saint Helens a stratovolcano or composite volcano. In contrast to the gently sloping shield volcanoes of Hawai'i, which typically erupt nonexplosively and produce lavas that can flow great distances, composite volcanoes tend to erupt explosively, like the 1980 eruption of Mount Saint Helens that shot ash sixty thousand feet into the air.

Mount Adams and Mount Saint Helens are part of the chain of fifteen active volcanoes in the Cascade Range—extending from Mount Garibaldi in British Columbia to Lassen Peak in northern California.

• • •

PAGE 44: SUMMIT OF WIZARD ISLAND, CRATER LAKE NATIONAL PARK, OREGON, 1992

A native Klamath legend describes the chief of the Below World, named Llao, who sat on his throne above Crater Lake and ruled over a race of giant crayfish that populated the cold waters below. Llao's archenemy, Skell, chief of the Upper World, captured and dismembered him, then threw the pieces into the lake. The crayfish ate Llao's body parts, not realizing they were devouring their own king. When they discovered the truth, they wept profusely, filling Crater Lake with tears and raising the water to its present level. Llao's head then surfaced, becoming Wizard Island.

• • •

PAGE 45: WIZARD ISLAND OVERLOOK, CRATER LAKE NATIONAL PARK, OREGON, 1980

Before Mount Mazama blew its top seven thousand years ago, the peak was estimated to be as high as twelve thousand feet. A column of hot gas and magma forty-two times greater than that of the 1980 Mount Saint Helens eruption exploded into the air and fell as pumice and ash. As the magma chamber emptied, the underlying support weakened, and the walls of the volcano collapsed, forming a huge caldera.

Water from rain and snow filled the caldera to form Crater Lake, the deepest lake in the United States at 1,932 feet. A subsequent eruption created Wizard Island, a cinder cone that rises 764 feet above the water.

• • •

PAGE 47: "SHE WHO WATCHES," HORSETHIEF LAKE STATE PARK, WASHINGTON, 1994

There are numerous petroglyphs and pictographs on the smooth basalt cliffs of the Columbia River Gorge. To this day we know very little about the meaning of Indian rock art. Interpretations, however, are plentiful. One legend of She Who Watches recounts the story of Coyote paying a visit to an old woman Indian chief and changing her into a rock with the command, "Women will no longer be chiefs. . . . You shall stay here and watch over the people who live here."

Another possible interpretation is that She Who Watches represents a death-cult guardian spirit. The death cult arose in response to the epidemics that wiped out hundreds of thousands of Native American populations, for the tribes had no immunity to foreign diseases such as smallpox, tuberculosis, and measles. "Tsagiglalal," as the Wishram Indians call her, silently watches over the land of her vanished people.

• • •

PAGE 48: COLUMNAR BASALT, LATOURELL FALLS, COLUMBIA RIVER GORGE, OREGON, 1993

One of the most striking phenomena of the volcanic landscape is the architecture of columnar basalt. As lava cools, the limits of its tensile strength prevent it from cooling around a central core. Instead, it congeals around local centers of contraction. Surface cracks develop around the areas of shrinkage in a pattern of polygons. Conditions permitting, these cracks deepen to allow the formation of long columns.

• • •

PAGE 49: LATOURELL FALLS, COLUMBIA RIVER GORGE, OREGON, 1993

These basalt cliffs were created by lava flows that traveled from a fissure over four hundred miles away, creating the entire Columbia River Plateau. Latourell Falls is the second highest waterfall in the Columbia River Gorge, plunging 249 feet over cliffs stained with yellow lichen.

PAGE 50: TRIPLE FALLS,
COLUMBIA RIVER GORGE, OREGON, 1994

Triple Falls cascades over basalt cliffs in the Columbia River Gorge National Scenic Area, just east of Portland, Oregon. The hot spot source of the very fluid lava that long ago blanketed this area is now believed to be the cause of the thermal activity in Yellowstone National Park in western Wyoming.

• • •

PAGE 51: SOUTH FALLS,
SILVER FALLS STATE PARK, OREGON, 1995

The area encompassing Silver Falls State Park was created from soft volcanic ash deposits and millions of years of accumulated lava flows. Streams and waterfalls are diverted by the harder basaltic rock but erode the softer ash layers. The result is over ten waterfalls in the park. South Falls drops 177 feet over volcanic rocks.

• • •

PAGE 52: MALHEUR MAAR,
DIAMOND CRATERS, OREGON, 1994

Diamond Craters, in the southeast desert of Oregon, is known as "Oregon's Geological Gem." In addition to the more traditional lava flows and craters, there are maars, domes, tuff, bombs, spatter, and driblet spires. Malheur Maar is seven thousand years old and only six feet deep.

• • •

PAGE 53: JOHN DAY FOSSIL BEDS
NATIONAL MONUMENT, OREGON, 1993

These badlands are the eroded remnants of minerals and layers of volcanic ash from the ancestral Cascade Mountains. Over thirty million years of compaction, cementation, and recrystallization transformed the ash into hard and colorful rock.

• • •

PAGE 54: MOUNT WRANGELL, WRANGELL-SAINT
ELIAS NATIONAL PARK, ALASKA, 1995

Mount Wrangell, at 14,163 feet, is the tallest active volcano in Alaska and one of the largest shield volcanoes in the world. Although Mount Rainier in Washington is taller, Mount Wrangell has over three times Rainier's volume.

• • •

PAGE 55: VALLEY OF TEN THOUSAND SMOKES,
KATMAI NATIONAL PARK, ALASKA, 1995

The largest and most forceful volcanic eruption of this century occurred on June 6 and 7, 1912, from Novarupta on the Alaska Peninsula. That explosive event lasted sixty hours and covered a once-verdant valley with tuff that extended for fifteen miles in length, three to six miles in width, and in some places reached a depth of seven hundred feet.

In 1916, Robert Griggs, leader of a *National Geographic* expedition, surveyed the aftermath of the Novarupta explosion. Through the thick layers of volcanic ash deposits, small holes and cracks had developed, allowing gas and steam to escape from the heated water underground. Griggs named this hissing, steaming landscape the Valley of Ten Thousand Smokes. A geologist would have known that this phenomenon would not last, but Griggs, a botanist, did not, and successfully pressed for park preservation of the wondrous site. Today the valley is quiet, with only an occasional wisp of sulfurous smoke.

• • •

PAGES 56: MOUNT BLACKBURN, WRANGELL-SAINT
ELIAS NATIONAL PARK, ALASKA, 1995

At 16,390 feet, Mount Blackburn is the tallest peak in the Wrangell Mountains. It is the eroded remains of a much larger volcano. Study of this presumed shield volcano, with its collapsed summit caldera, is hampered by its extreme height, remoteness, and ice cover.

• • •

PAGE 57: ROOT GLACIER, WRANGELL-SAINT
ELIAS NATIONAL PARK, ALASKA, 1995

In the distance beyond the gray mountain called Doneho Peak is the snow-covered volcano Mount Blackburn. Long before Lieutenant Henry T. Allen of the U.S. Army gave it the name Blackburn, it was known to the Ahtna Athabaskan natives as K'als'i Tl'aadi, or "The One at Cold Waters."

• • •

PAGES 58–59: MOUNT BLACKBURN AND KENNICOTT GLACIER,
WRANGELL-SAINT ELIAS NATIONAL PARK, ALASKA, 1995

Wrangell-Saint Elias National Park encompasses 13.2 million acres. Five million acres are covered by glaciers and ice fields, with Mount Blackburn almost entirely blanketed by ice.

• • •

PAGE 60: AUGUSTINE VOLCANO, COOK INLET, ALASKA, 1995

Augustine Volcano's last eruption occurred in 1986. It began explosively, with ash plumes rising more than six miles above the vent and pyroclastic flows racing down the flanks of the volcano. Later, lava began erupting near the summit, adding eighty-two feet to the existing lava dome.

Early recorded history of Augustine Volcano shows eruptions spaced at seventy-year intervals. More recently, the intervals have shortened to ten years.

• • •

PAGE 61: AUGUSTINE VOLCANO, COOK INLET, ALASKA, 1995

See notes for pages 60 and 63.

• • •

PAGE 63: AUGUSTINE VOLCANO, COOK INLET, ALASKA, 1995

Augustine Volcano is seen here with a petite puff of smoke, rising peacefully out of the Cook Inlet. Appearances are deceptive, for Augustine is one of the most active and explosive volcanoes in the inlet. Of special concern are the debris avalanches that can occur during eruptions. Once the rocks slide off the volcano's flanks and into the sea, they can trigger tsunamis. Since Cook Inlet is home to 60 percent of Alaska's population, this volcano is carefully monitored by the Alaska Volcano Observatory.

California and the Southwest

California and the Southwest cover over half a million square miles of land, which includes the four great American deserts: the Great Basin, the Sonoran, the Mojave, and the Chihuahuan. This region brings to mind beautiful red rock spires and buttes, the Grand Canyon, and vast expanses of desert—but not volcanoes.

Perhaps this is because, with the exception of Mount Lassen's major eruption in 1914–1917, there has been no volcanic activity in this region during the twentieth century. Although the numerous volcanic fields here have remained quiet, this does not mean they will not erupt again. The list of active and potentially active volcanic sites prepared by the United States Geological Survey includes California's Long Valley Caldera, Arizona's San Francisco Volcanic Field, and New Mexico's Bandera Field, among others. Over fifteen national parks and monuments in this section of the western United States contain volcanic features.

• • •

PAGE 65: SP CRATER,
SAN FRANCISCO VOLCANIC FIELD, ARIZONA, 1995

Of the 550 vents that make up the San Francisco Volcanic Field in northern Arizona, SP Crater is the only one known solely by initials. Early cartographers purportedly found its full name, Shit Pot, too embarrassing to print on maps.

• • •

PAGE 66: GRIMES POINT ARCHAEOLOGICAL AREA, NEVADA, 1993

Grimes Point in the Great Basin Desert of western Nevada was once under water. Up until the end of the Pleistocene Ice Age about ten thousand years ago, this region was covered by Lake Lahontan.

Grimes Point is an area abundant with petroglyphs. Once the smooth surfaces of basalt lava boulders were etched, the carving revealed lighter subsurfaces. There is no way to date rock art accurately, nor to decipher the meaning of the images.

• • •

PAGE 67: SHIP ROCK, NEW MEXICO, 1995

Ship Rock got its name from white people who saw the dark volcanic monument and thought it resembled a ship with sails. The Navajos, however, liken it to a bird and call it Tsé'bĭtaï, or "Winged Rock."

In one Navajo legend, a great warrior named Nayénĕzgani traveled to Winged Rock in search of Tse'nă'hale, a winged monster. Nayénĕzgani was captured in the talons of the creature and dropped on a ledge far below. This fall had killed all previous intruders of Tse'nă'hale, but not Nayénĕzgani. He was saved by two possessions: the life feather, a gift of Spider Woman, and a bag filled with blood that he was given as a trophy for a previous slaying. When the warrior fell, he cut open the bag and the blood spilled out, tricking the monster into thinking he was dead. This allowed the hero to escape and later slay the monster.

The legend also refers to two great snakes turned to stone upon Nayénĕzgani's visit to Winged Rock. The dikes radiating out from the sides of Ship Rock coincide with this account.

PAGE 69: SHIP ROCK, NEW MEXICO, 1988

Ship Rock is a volcanic neck, the so-called innards of a volcano or the volcanic "plumbing system." It is the remains of an ancient volcano in the New Mexican desert that never broke ground but hardened in a fissure far below the surface of the earth. Erosion has worn away the surrounding hills and left a monument seventeen hundred feet above ground. Pictured here is a smaller fragment of one of Ship Rock's many protrusions, as well as one of the dikes.

• • •

PAGES 70–71: FERN CAVE,
LAVA BEDS NATIONAL MONUMENT, CALIFORNIA, 1994

Lava Beds National Monument in northern California is an area of extensive lava flows, cinder cones, and, most remarkably, almost two hundred lava tubes. The conditions for ferns to grow in the middle of the desert would be, at best, adverse—the closest fern is probably one hundred fifty miles away on the moist Pacific coast. Yet once the seed was deposited (probably by wind or bird), the cool and damp interior of the cavelike tube provided a perfect growing environment. Fern Cave is a sacred spot to the Modoc Indians.

• • •

PAGE 72: DEVILS POSTPILE
NATIONAL MONUMENT, CALIFORNIA, 1994

Among the premier places in the world to view columnar basalt—including the Giant's Causeway in Ireland and Fingal's Cave in Scotland—is Devils Postpile, on the western slope of the Sierra Nevada in California. About one hundred thousand years ago, lava cooled and formed surface cracks that hardened into patterns and deepened, resulting in the formation of sixty-foot columns. Subsequent erosion caused fragments of rock and column to pile up below the cliff.

• • •

PAGE 73: BLACK POINT,
MONO LAKE TUFA STATE RESERVE, CALIFORNIA, 1992

About thirteen thousand years ago, Black Point was created by a volcanic eruption that took place under water. Mono Lake sits at the northernmost point of the Long Valley Caldera, a huge crater that stretches twenty-five miles to the south. Exactly one week after the May 1980 eruption of Mount Saint Helens in Washington, a strong earthquake swarm at Long Valley caught the attention of geologists. Since then the caldera has lifted two feet, often a signal that magma is rising to the surface and pushing up the earth. This, combined with the earthquake swarms, suggests that a volcanic eruption in the near future is possible.

• • •

PAGES 74–75: APPROACHING STORM AT DAWN,
MONO LAKE TUFA STATE RESERVE, CALIFORNIA, 1991

Mono Lake, the second oldest lake in North America, is surrounded by volcanoes that began erupting thirteen million years ago. Thousands of feet of volcanic ash, deposited in the lake over centuries, keep the water alkaline, creating a favorable environment for calcium carbonate to precipitate out of the hot springs, forming the tufa spires.

Mark Twain called Mono Lake a "solemn, silent, sailless sea . . . lonely tenant on the loneliest spot on earth."

• • •

PAGE 77: LITTLE UBEHEBE CRATER,
DEATH VALLEY NATIONAL PARK, CALIFORNIA, 1991

Little Ubehebe Crater is part of a volcanic field of maar craters. The crater was probably formed by the reaction of lava with water-saturated earth, resulting in the explosive ejection of debris.

• • •

PAGE 78: SNOW CANYON STATE PARK, UTAH, 1993

Southwest Utah was once an area active with volcanism. Snow Canyon is surrounded by extinct volcanoes to the north and Lava Point, in Zion National Park, to the east. Nearby volcanoes dumped ash and lava into Snow Canyon beginning three million years ago and continuing until one thousand to two thousand years ago.

In 1953, Snow Canyon received a different kind of explosive fallout in its location downwind of the Upshot-Knothole series of eleven atmospheric nuclear tests in the Nevada desert. The stark volcanic landscape mixed with red sandstone made this canyon a popular location for Hollywood Westerns. During the filming of *The Conqueror* in 1954, Snow Canyon was still so radioactive that Geiger counters ticked off the scale. To date, 91 of the 220 actors and crew that director Dick Powell brought in for the chase and battle scenes have died of cancer and cancer-related diseases, including John Wayne, Susan Hayward, Agnes Moorehead, and the director himself.

• • •

PAGE 79: CITADEL RUIN,
WUPATKI NATIONAL MONUMENT, ARIZONA, 1992

This structure, now known as Citadel Ruin, was once a two-story pueblo containing approximately thirty rooms. The masonry incorporates lava with the sandstone. It sits on a lava butte with a commanding view of the surrounding land. In the distance, one sees a line of cinder cones that are part of the San Francisco Volcanic Field. Although no eruption has occurred for over seven hundred years, the United States Geological Survey still lists this area as a potentially active volcanic site.

• • •

PAGE 81: BENTONITE HILLS,
CAPITOL REEF NATIONAL PARK, UTAH, 1988

The Bentonite Hills are made up of mud, silt, sand, and volcanic ash deposited in swamps and lakes one hundred forty million years ago during the Jurassic era. The erosion of this soft, claylike substance often reveals petrified wood, dinosaur bones, and lava.

• • •

PAGE 82: DONEY MOUNTAIN,
WUPATKI NATIONAL MONUMENT, ARIZONA, 1992

The volcanic formation referred to as Doney Mountain is actually four elongated craters built along a fault line. In the distance lies the Painted Desert, made up of multihued layers of soft clay, minerals, and volcanic ash.

PAGE 83: TUFF CANYON OVERLOOK,
BIG BEND NATIONAL PARK, TEXAS, 1990

Indian legend says that after making the Earth, the Great Spirit simply dumped all the leftover rocks on the Big Bend. Geologists humorously refer to this conglomeration as an "eggbeater formation." Tuff, or hardened volcanic ash, was deposited into this canyon by a volcano in the nearby Chisos Mountains.

• • •

PAGE 84: SUNSET CRATER
NATIONAL MONUMENT, ARIZONA, 1994

Sunset Crater is the youngest volcano in the San Francisco Volcanic Field. It began erupting in the winter of 1064–1065 and continued for another two hundred years. During the course of its development, it expelled an estimated one million tons of material. Most of the cinder fell back around the vent, creating the thousand-foot cone, but prevailing winds also carried the ash and cinder over eight hundred square miles.

The impact of this fiery event is interwoven into many Hopi legends. In the Ka'naskatsina legend, the creation of Sunset Crater plays a pivotal role in a magical story involving betrayal and revenge. Even after seven centuries of oral transmission, the story contains accurate details of the dramatic eruption.

• • •

PAGE 85: BLACK BUTTE, SHASTA-TRINITY
NATIONAL FOREST, CALIFORNIA, 1993

Black Butte is a steep plug dome volcano that rises 6,325 feet just west of Mount Shasta in northern California. Four successive eruptions from the base of Mount Shasta squeezed a thick, cookie-dough-like material from this vent, forming the butte.

• • •

PAGE 86: GRAND FALLS OF THE
LITTLE COLORADO RIVER, ARIZONA, 1994

Approximately one hundred thousand years ago, a cinder cone erupted ten miles southeast of the present location of Grand Falls. As lava filled the canyon, it displaced the river from its gorge, diverting it around the lava flow and over the cliffs, creating the 185-foot terraced waterfalls.

• • •

PAGE 87: SUPERSTITION MOUNTAINS,
LOST DUTCHMAN STATE PARK, ARIZONA, 1995

Thick layers of volcanic rocks make up the Superstition Mountains. Deposits of pyroclastic flows, lava domes, lahars, and lava flows from seventeen to twenty-five million years ago testify to a period of intense volcanism in what is now the Sonoran Desert of southern Arizona.

• • •

PAGE 89: MOUNT LASSEN AND THE FANTASTIC LAVA BEDS,
LASSEN VOLCANIC NATIONAL PARK, CALIFORNIA, 1994

Native people describe Mount Lassen by many names: Broken Mountain, Mountain Ripped Apart, Little Shasta, and Fire Mountain. Before Mount Saint Helens erupted in 1980, Lassen Peak was the most recent volcanic eruption in the lower forty-eight. It is the southernmost volcano of the active Cascade Range, which extends north into British Columbia.

Lassen Peak probably first erupted through a vent on Mount Tehama's flank. The remnants of Mount Tehama's caldera are seen in Brokeoff Mountain, Mount Diller, Pilot Pinnacle, and Mount Conard, which would have made the base of Mount Tehama more than eleven miles wide. Mount Lassen, at 10,457 feet, is considered the world's largest plug dome volcano.

• • •

PAGE 90: MAMMATUS CLOUDS, CINDER CONE,
LASSEN VOLCANIC NATIONAL PARK, CALIFORNIA, 1992

Spectacular cloud formations like this one, known as mammatus or mammatiform clouds, are often seen before or after severe thunderstorms. Characteristic of this phenomenon are the rounded protuberances that hang down from the undersurface of a thunderhead's anvil.

The volcanic ash in the vicinity of Cinder Cone is quite thick, and water percolates through it quickly. This factor, combined with a short growing season, makes it difficult for plants to take hold in this environment. Because of the porous cinders, water has little erosional effect on the cone. The volcano's shape will be preserved, assuming there are no new eruptions.

• • •

PAGE 91: CINDER CONE,
LASSEN VOLCANIC NATIONAL PARK, CALIFORNIA, 1992

All volcanoes are born from vents. Magma either flows outward or, if large quantities of gas and steam are present, erupts explosively upward in fragments that fall back around the vent to build a cone. Cinder Cone is a classic example of a volcano known as a tephra cone. Successive explosions of cinder, ash, and bombs created the six-hundred-foot cone.

• • •

PAGES 92–93: SUMMIT OF CINDER CONE,
LASSEN VOLCANIC NATIONAL PARK, CALIFORNIA, 1992

Those who reach the summit of Cinder Cone and get a chance to study the crater will notice an unusual phenomenon—there are two and possibly four craters instead of the usual one, suggesting separate eruptions. Twice we have hiked to the summit and did not have time to study this feature—on both trips we were overtaken by sudden, violent thunderstorms. When you are on the tallest peak for miles around during a thunderstorm, your metal tripod is unfortunately an excellent electrical conductor.

Yellowstone and the Northern Mountain States

To this day, numerous theories exist about the thermal activity and the origins of the Yellowstone Basin. One theory, still hotly contested, is that the great lava eruption in southeastern Oregon that created the Columbia River Plateau was caused by a giant meteorite that pierced the crust of the earth. The size of such a meteorite is difficult to fathom when we take into account that the average thickness of the earth's crust is twenty-five miles, but according to the theory, once the crust was broken by the meteorite, a magma reservoir formed, creating a hot spot. As the continental plate moved west, volcanic eruptions occurred to the east, in the Snake River Plateau and Craters of the Moon in Idaho, and Yellowstone in Wyoming.

Also contested once but now almost universally endorsed is the theory that a single source (an active magma chamber) is heating all the geysers and thermal features of the Yellowstone Basin, and that Yellowstone is one of the hottest spots on earth.

The evidence came in numerous stages. First a Harvard graduate student named Francis Boyd proposed in his thesis of 1950 that the vast layers of welded tuff were deposited by cataclysmic pyroclastic flows. Boyd went on to prove that the eruption that produced these flows had occurred more recently than was previously suspected. The caldera formed by such a tremendous explosion was so large that several generations of geologists had not seen it, even though they were standing right inside of it the whole time.

Then, in 1959, a 7.5 magnitude earthquake on the park's western border set off 298 geysers and thermal features at the same time. The temperature of the hydrothermal features in Yellowstone rose five and a half degrees while water levels dropped in wells as far away as New York and California. The event brought scientists swarming to the park to conduct studies.

What we can now deduce from their research is that the earth's crust under Yellowstone is only two to six miles thick and is resting on a magma reservoir, demonstrating that the thermal force heating the geysers is still active. The heat is so intense that geologists believe it could not possibly be the cooling from the last eruption, but instead represents the beginnings of a new eruptive cycle. With more sophisticated mapping techniques we also know that there were actually three cataclysmic eruptions here about six hundred thousand years apart—with the last eruption taking place approximately six hundred thousand years ago. Twin domes are now rising within the ancient caldera, another possible signal that a new eruptive cycle is underway.

• • •

PAGE 95: DEVILS TOWER NATIONAL
MONUMENT, WYOMING, 1994

Devils Tower is a volcanic plug, the hardened inner core of a magma chamber. Geologists believe this plug did not surface above the earth until the softer material surrounding it wore away. After sixty million years of erosion, what remains is an 867-foot protrusion. The ridges formed when magma forced its way into overlying sedimentary rock and cooled underground. This process of lava cooling into polygonal columns is known as columnar jointing.

The local Kiowa Indians recount a legend about Devils Tower and the forming of the ridges. One version had eight children—seven sisters and their brother—at play, when the boy was suddenly transformed into a bear. The sisters became terrified and began to run, and the bear chased them until they reached the stump of a great tree. The tree stump spoke and bade the sisters to climb upon it, and as they did, the stump rose into the air, just out of the bear's reach. He reared against the trunk and clawed the bark, scoring the ridges. The ridged structure became Devils Tower, and the seven sisters were borne into the sky and became the stars of the Big Dipper.

The uniqueness of Devils Tower not only captured Native Americans' imaginations, but also Hollywood's. In Steven Spielberg's *Close Encounters of the Third Kind,* Devils Tower serves as the symbol of and setting for the alien encounter.

• • •

PAGE 96: SPATTER CONE, CRATERS OF THE MOON NATIONAL MONUMENT, IDAHO, 1991

The volcanic formations that make up the eighty-three-square-mile area known as Craters of the Moon were created by a fissure-vent eruption that began fifteen thousand years ago and continued for the next thirteen thousand years.

A spatter cone forms during the later stages of an eruptive episode. In the early stages, gas-rich lava shoots high into the air, creating cinder cones. However, when the lava is gas depleted, it oozes a thick, pasty substance that builds up around the vent, forming the steep-walled spatter cone.

• • •

PAGE 97: HOT SPRINGS STATE PARK, THERMOPOLIS, WYOMING, 1994

The name Thermopolis is a compound of Greek words meaning "city of heat." This hot city is home to the world's largest mineral hot spring, heated by the volcanic rocks that lie below.

• • •

PAGE 98: MINERVA SPRING, MAMMOTH HOT SPRINGS, YELLOWSTONE NATIONAL PARK, WYOMING, 1990

The steplike terraces of Mammoth Hot Springs are created by hot water carrying minerals that percolate to the surface and form calcium carbonate that precipitates as travertine. At Mammoth Hot Springs, more than two tons of travertine accumulate in just one day. New calcium carbonate is white and grays with age, but as water cascades from terrace to terrace, the water cools, allowing algae to grow in shades of green, yellow, red, and orange.

• • •

PAGE 99: HOT SPRINGS STATE PARK, THERMOPOLIS, WYOMING, 1994

This hot spring, by the same terrace-building processes at work in Yellowstone, creates spectacular lime and gypsum formations.

• • •

PAGES 100–101: WHEELER GEOLOGIC AREA, SAN JUAN MOUNTAINS, COLORADO, 1995

The formations that make up the Wheeler Geologic Area are eroded volcanic tuff. The tuff was deposited sixty-five million years ago in a violent eruption of the La Garita Caldera, which spread thousands of tons of volcanic ash over the central San Juan Mountains. This material is highly erosive when exposed to the elements.

• • •

PAGE 102: OPALESCENT POOL, YELLOWSTONE NATIONAL PARK, WYOMING, 1991

Black Sand Basin, part of the Upper Geyser Basin, was named for the black obsidian sand found in this group of geysers, pools, and springs. Opalescent Pool, though only six feet deep, flooded and killed most of the trees in this basin in the 1950s. Since then, silica has precipitated up the trunks of the lodgepole pines, creating the white bottoms, thus the name "bobby sock" trees. In time, the silica will impregnate and petrify the wood.

• • •

PAGE 103: PUNCHBOWL SPRING RUNOFF, YELLOWSTONE NATIONAL PARK, WYOMING, 1991

In Yellowstone, the colors seen in the hot springs and areas of geyser runoff are generally caused by the pigmentation of algae. Different plant communities grow within distinct ranges of water temperatures. Nearest to the source, the water temperature is 199 degrees Fahrenheit (boiling point at this elevation of 7,300 feet). Very few organisms can live in that range. Here in the Upper Geyser Basin, in the runoff from Punchbowl Spring, cyanobacteria live, creating the green and orange ribbons of color.

• • •

PAGE 104: PORCELAIN BASIN, YELLOWSTONE NATIONAL PARK, WYOMING, 1992

Porcelain and Back Basins comprise the Norris Geyser Basin, which is the hottest spot in Yellowstone National Park. In 1929, the Carnegie Institute conducted tests of subsurface temperatures but had to abandon testing in the Norris Basin because the drilling rig was in danger of melting. At a depth of 265 feet, the temperature reached an astounding 401 degrees Fahrenheit. The stark appearance of Porcelain Basin is due to the acidic environment, which is not conducive to plant and algae growth.

• • •

PAGE 105: EXCELSIOR SPRING RUNOFF, YELLOWSTONE NATIONAL PARK, WYOMING, 1990

The thermal runoff in this photograph is from the now-dormant Excelsior Geyser in the Midway Geyser Basin. At the time of its peak activity in the 1880s, Excelsior was considered one of the largest geysers in the world, spouting to a height of three hundred feet. Now classified as a hot spring, it discharges an incredible 4,050 gallons of water per minute.

• • •

PAGE 106: BLACK SAND POOL, YELLOWSTONE NATIONAL PARK, WYOMING, 1991

It is estimated that nearly ten thousand thermal features exist in Yellowstone. Geysers are one type of hot spring requiring a nearly vertical underground tube for eruptions to take place. When underground networks allow superheated water to cool substantially before reaching the surface, it bubbles up from the vent as a hot spring.

• • •

PAGE 107: LION GROUP GEYSER ERUPTING, YELLOWSTONE NATIONAL PARK, WYOMING, 1990

Geysers are formed when water seeps through cracks and fissures and is heated by rock, which in turn is heated by magma. The water reaches a temperature above its surface boiling point, but underground pressure prevents it from boiling, as in a pressure cooker. As the water rises to the surface, the pressure is released, and boiling explosions shoot steam into the air in the form of a geyser.

The Lion Group is composed of four different geysers connected subterraneanly. Some claim it is named for the roar that can be heard before an eruption, but it was actually named in 1881 for its resemblance to the body and maned head of a reclining lion.

• • •

PAGE 109: PATH THROUGH PORCELAIN BASIN,
YELLOWSTONE NATIONAL PARK, WYOMING, 1992

After enduring the crowds of Yellowstone National Park in summer, it is hard to imagine that in winter you and the wildlife can have the park to yourselves. The steam coming off the thermal features is more pronounced in the frigid temperatures of Wyoming in January.

We left the park entrance at sunrise by snowmobile en route to the Norris Geyser Basin, the hottest spot in the park. The trip took forty-five minutes at an average speed of about forty miles per hour in ten-degree-Fahrenheit temperature. We do not want to hazard a guess about windchill but we were sure we had done irreversible damage to our toes and fingers.

However, when we arrived at our destination and observed the winter wonderland shrouded in mist and fog, we immediately forgot our pain and grabbed our cameras. For the first time ever, our shutters froze, and the frustration of not being able to record what we were seeing was monumental. A half hour later, after walking through the steam and warming our cameras under our down coats, we had thawed the delicate mechanisms.

• • •

PAGE 111: INFERNO CONE, CRATERS OF
THE MOON NATIONAL MONUMENT, IDAHO, 1991

The Apollo astronauts preceded our visits to many of the volcanic national parks and monuments. NASA, in its search for practice grounds for the first moon landing, sent Neil Armstrong and crew to lava fields and craters from Iceland to Hawai'i. Not quite believing all the placards that we happened upon in volcanic parks commemorating their training grounds, at NASA headquarters we found documentation of the astronauts hiking through Kīlauea and the Valley of Ten Thousand Smokes, as well as the earthly Craters of the Moon.

• • •

PAGES 112–113: PORCELAIN BASIN,
YELLOWSTONE NATIONAL PARK, WYOMING, 1992

The larger thermal pools in the Norris Geyser Basin radiate with brilliant colors from blue to emerald green. The pools change hue from day to day depending on atmospheric conditions and the amount of particulate matter in the water.

• • •

PAGE 115: BACK BASIN, YELLOWSTONE
NATIONAL PARK, WYOMING, 1992

The Norris Geyser Basin, on a calm winter morning, creates its own fog bank due to the heat generated by the thermal features reacting with the frigid air. Unlike Porcelain Basin, which is barren, Back Basin supports plant life and trees—one of which is pictured here, buried under a thick blanket of snow.

PAGE 116: ECHINUS GEYSER ERUPTING,
YELLOWSTONE NATIONAL PARK, WYOMING, 1990

The name of this geyser comes from the Greek word for sea urchin. Apparently the spine-covered crater reminded early visitors of that creature. In the Norris Geyser Basin, Echinus is the largest predictable geyser, erupting to a height of fifty to one hundred feet for up to sixty minutes. These two men had plenty of time to sit and watch water evaporate.

• • •

PAGE 117: ECHINUS GEYSER ERUPTING,
YELLOWSTONE NATIONAL PARK, WYOMING, 1990

In Back Basin of the Norris Geyser Basin, two geysers lie close to each other, yet are very different in character. Steamboat Geyser is the largest geyser in the world, reaching heights of 380 feet. However, it is one of the most erratic spouters, with periods of dormancy that have lasted for decades. Echinus Geyser, on the other hand, rivals Old Faithful for consistency, shooting water into the air every thirty to seventy-five minutes.

• • •

PAGE 119: FISHING CONE,
YELLOWSTONE NATIONAL PARK, WYOMING, 1990

This thermal cone, located in the West Thumb Geyser Basin, received its name from early explorers who caught their fish in Yellowstone Lake, then tossed the fish, while still on the line, into the Fishing Cone—simmering their catch.

GLOSSARY

'A'Ā: Hawaiian word used to describe a lava flow characterized by a rough, jagged, fragmented surface.

ACTIVE VOLCANO: A volcano that is currently erupting or has erupted in recorded history.

ASH: Fine particles of pulverized rock blown from a volcano.

BASALT: A dark-colored, fine-grained lava. The most common of earth's volcanic rocks, basaltic lavas make up all the ocean floors and many continental formations. Often very fluid, basaltic lavas can flow great distances from their sources, forming broad lava plains such as the Columbia River Plateau.

BOMB, VOLCANIC: A still-viscous lava fragment ejected in an explosive eruption that takes on a rounded shape while in flight.

CALDERA: A large circular or basin-shaped depression, usually at the summit of a volcano. Often, a caldera forms after a volcanic explosion because the support has given way after the magma chamber has emptied or drained away.

CINDER CONE: A steep conical hill formed when cinders and other loose material propelled into the air by a volcanic eruption fall back around the vent.

COLUMNAR BASALT (COLUMNAR JOINTING): The process of basaltic rock fracturing into polygonal columns that are characteristic of cooling lava.

COMPOSITE VOLCANO: See Stratovolcano.

CONTINENTAL PLATE: See Plate tectonics.

CRATER: A steep-walled depression usually at the summit of a volcanic cone, situated above the vent that feeds the volcano and from which volcanic materials are usually ejected.

DIKE: A sheetlike vertical intrusion of molten rock that has forced its way through an existing rock structure.

DOME: A mushroom-shaped cap formed when lava too viscous to flow during an eruption piles over the vent.

DORMANT VOLCANO: A once-active volcano that is now inactive or "sleeping." Dormant volcanoes are considered capable of future eruptions.

DRIBLET SPIRE: Similar to spatter, these are formations created by thick globs of molten magma being ejected a few feet into the air and falling back down to stack up over the vent.

EARTHQUAKE, VOLCANIC: A localized and light shaking of the ground caused by rock fracturing as magma rises to the surface of the earth.

EXTINCT VOLCANO: A "dead" volcano that is not currently erupting and is not expected to do so in the future.

FAULT: A fracture or crack associated with movement in the earth's crust. Movement along a fault can cause either earthquakes or volcanoes.

FISSURE: A fracture in the earth's crust.

FLANK ERUPTION: An eruption on the side of a volcano rather than at the summit.

FUMAROLE: A vent or opening in the earth's surface that emits steam or volcanic gases.

GEYSER: A natural hot spring that intermittently shoots a column of superheated water and steam into the air.

GLACIER: An ice mass or ice river formed by the compaction of snow that flows down a mountain by its own weight and gravity.

HAWAIIAN ERUPTION: Effusive eruptions of basaltic lava typical of the Hawaiian shield volcanoes. Usually nonexplosive, Hawaiian eruptive activity commonly produces lava fountaining, lava lakes, and huge quantities of fluid lava.

HOT SPOT: A persistent heat source in the upper mantle of the earth's crust unrelated to plate boundaries where molten rock is continually rising. Isolated hot spots underlie the Hawaiian Islands and the Yellowstone region.

HOT SPRING: A natural spring where water above ninety-eight degrees Fahrenheit is discharged. Hot springs are closely related to geysers except they do not discharge water and steam into the air.

LAHAR: An Indonesian term for a mudflow or a landslide of volcanic debris flowing down the flank of a volcano.

LAVA: Magma that reaches the earth's surface through a volcanic eruption or a fissure in the earth's surface. The term is most commonly applied to streams of molten rock flowing from a volcanic vent. It also refers to solidified volcanic rock.

LAVA FOUNTAIN: A spray of molten rock shooting into the air from a volcanic vent or fissure.

LAVA TUBE: Created when the outer crust of a lava stream cools and solidifies, but the molten interior continues flowing. When the eruption ceases, and the molten interior has drained away, a cavelike tunnel or tube remains.

MAAR: Shallow, flat-floored craters left behind after a volcanic eruption in which no magma surfaces during the explosion.

MAGMA: Gas-rich molten rock beneath the earth's surface. Magma that reaches the earth's surface is called lava.

MANTLE: The intermediate layer of the earth's interior that lies between the molten core and the outer crust. It is within this region of hot plastic rock that magma is generated.

NECK: The lava-filled conduit of a former volcanic vent, exposed by the erosion of softer surrounding materials.

OBSIDIAN: A dense, often black volcanic glass almost devoid of bubbles or any mineral crystals.

PĀHOEHOE: Hawaiian word denoting a basaltic lava with a smooth, ropy, wrinkled, or billowy surface.

PLATE TECTONICS: An important geologic theory according to which the earth's crust is broken into approximately a dozen slabs or plates that slowly move about the earth's surface, causing faulting, volcanism, earthquakes, and mountain building along their margins.

PLUG: Solidified magma that fills the conduit of a volcano. Often it is harder and more resistant to erosion than the surrounding composite materials of the cone and can remain as a solitary pinnacle or neck when the rest of the volcano has been worn away.

PLUG DOME VOLCANO: A volcanic dome generally formed by highly viscous magma containing insufficient gas concentrations to cause violent explosions; instead the magma is squeezed up through the vent.

PLUME: A column of magma rising from deep within the mantle responsible for hot spot volcanoes. The term also refers to a column of steam or other gases rising from an erupting volcano.

PUMICE: A form of highly porous volcanic glass caused by violent eruptions. Pumice is so filled with gas-bubble holes that it resembles a sponge and is usually light enough to float on water.

PYROCLASTIC FLOW: An avalanche of fast-moving incandescent rock fragments, hot volcanic ash, and gas that travels downslope like a heavy fluid.

RIFT SYSTEM: The area along the boundaries of separating plates. Magma erupts along active rifts where new crust is being created.

RIFT ZONE: A linear belt along the flank of a volcano where eruptions can occur.

RING OF FIRE: The popular term for a region of intense seismicity and volcanism that surrounds the Pacific Ocean basin. This area contains approximately 75 percent of the world's active volcanoes.

SHIELD VOLCANO: A broad, gently sloping volcano in the shape of a shield with the curved side up, formed by the accumulation of numerous fluid lava flows.

SPATTER: Thick, pasty fragments of lava.

STRATOVOLCANO: A steep volcanic cone made up of alternating layers (strata) of lava flows and pyroclastic material. Also known as a composite cone or volcano.

SUBDUCTION ZONE: The zone of convergence of two tectonic plates, one of which sinks beneath the other. The Pacific plate descending beneath the North American plate is a classic example and is responsible for the numerous volcanoes in Alaska and the active Cascade Range in the Pacific Northwest.

TUFA: (Nonvolcanic) sedimentary rock made up of calcium carbonate or silica and formed from percolating groundwater or from solution in a lake or spring.

TUFF: A volcanic rock made up of generally fine-grained ash that is either compacted or welded by the heat of an eruption, as in welded tuff.

VENT: An opening in the earth's surface through which volcanic material is ejected.

VOLCANO: A vent in the surface of the earth through which magma erupts. Also the resulting landform produced by the erupted material.

INDEX OF PHOTOGRAPHS

Alaska

Arizona

California

Colorado

Hawai'i

Idaho

Nevada

New Mexico

Oregon

Texas

Utah

Washington

Wyoming

ACKNOWLEDGMENTS

During the course of this project, many friends and colleagues offered encouragement, advice, and much-needed criticism. To the following, we are most grateful: Anthony Accardi, Susan Bell, Niki and Peter Berg, Rebecca Busselle, Anna Gay Del Vescovo, Mitch Epstein, Joan Feeney, David Feingold, Christoph Gielen, Carey and Pepe Karmel, Marcia Lippman, Bruce Phillips, Matt Postal, Yancey Richardson, Michael Roukes, Janet Russek, David Scheinbaum, Sage Sohier, Eliot Weinberger, Jacob Weisberg, and Larry Wolfson. Special "aloha" to David Ulrich.

We are also grateful to the photo editors who gave us assignments to the volcanic lands we so love. Thanks are extended to Kathleen Klech and Dana Nelson at *Condé Nast Traveler,* Deborah Needleman, formerly of *Men's Journal,* Nina Subin of the *New York Times' Sophisticated Traveler,* Hazel Hammond, formerly of *Travel & Leisure,* and Bill Black and Stephanie Syrop, formerly of *Travel Holiday.*

Special thanks to David Clague, former director of the Hawaiian Volcanoes Observatory, for an exhilarating "hot" walk on lava. There are many scientists, too numerous to name, within the National Park Service and the United States Geological Survey who steered us in the right direction to sites and generously provided us with current research— we are most grateful.

Many of the locations we photographed were best viewed from the air; we are indebted to the pilots who brought us back safely. None of the aerial photographs would have been possible without the aesthetic and life-saving insights that Russell Munson so graciously shared with us.

We owe much to our friend and superb designer Louise Fili and her assistant Tonya Hudson, who brought extraordinary and imaginative design solutions to this project. To our agent, Bob Markel, thanks for the necessary assistance.

To our editor, Janet Swan Bush, and her most able assistant, Janice O'Leary, as well as Betsy Uhrig and the entire Bulfinch team who worked on *Hot Spots,* we wish to express special thanks for their guidance, patience, and remarkable skills in seeing this project through to publication. Thanks also to Brian Hotchkiss, formerly of Bulfinch Press, for believing in this project.

To all, thank you for guiding us on this journey.